JOHN DEERE MODEL B

PHOTO ARCHIVE

JOHN DEERE MODEL B
PHOTO ARCHIVE

Photographs from the
Deere & Company Archives

Edited with introduction by
P. A. Letourneau

Iconografix
Photo Archive Series

Iconografix
P.O. Box 18433
Minneapolis, Minnesota 55418 USA

Revised July 1993

Library of Congress Card Number 92-74595

ISBN 1-882256-01-8

93 94 95 96 97 98 5 4 3 2

Cover and book design by Lou Gordon

Printed in the United States of America

PREFACE

The histories of machines and mechanical gadgets are contained in the books, journals, correspondence and personal papers stored in libraries and archives throughout the world. Written in tens of languages, covering thousands of subjects, the stories are recorded in millions of words.

Words are powerful. Yet, the impact of a single image, a photograph or an illustration, often relates more than dozens of pages of text. Fortunately, many of the libraries and archives that house the words also preserve the images.

In the *Photo Archive Series*, Iconografix reproduces photographs and illustrations selected from public and private collections. The images are chosen to tell a story...to capture the character of their subject. Reproduced as found, they are accompanied by the captions made available by the archive.

The Iconografix *Photo Archive Series* is dedicated to young and old alike, the enthusiast, the collector and anyone who, like us, is fascinated by "things" mechanical.

ACKNOWLEDGMENTS

The photographs and illustrations which appear in this book were made available by the Deere & Company Archives. We are most grateful to the company and to Les Stegh, Deere Archivist.

Cross-Sectional View of John Deere
Model "B" Tractor

INTRODUCTION

In 1928, Deere & Company introduced its first general purpose or row-crop tractor, the Model GP. Deere followed with the General Purpose Model A, in 1934.

Despite its widespread acceptance, the Model A was too large for many farmers. In the early 1930s, more than half of America's farms were smaller than 100 acres. It was the potential for a smaller row-crop tractor, suited to the needs of this majority, that inspired Deere to develop the Model B. Introduced in 1935, the Model B was an immediate and long-lived success. More than 300,000 units were sold by the end of its production run in 1952.

The B Series was built in 11 distinct configurations: the 2-wheel, tricycle front Model B; the single front wheel Model BN; the adjustable front axle Model BW; the standard axle Model BR; the grove and orchard Model BO; the high-clearance versions of the BN and BW (Model BNH and Model BWH); the narrow rear axle versions of the BW and BWH (Model BW-40 and Model BWH-40); the industrial tractor Model BI; and the John Deere-Lindeman crawler tractor.

The Model B offered all the features of the Model A, but built on a smaller scale: a rugged, horizontal 2-cylinder motor that generated 14.30 brake horsepower and 9.38 drawbar horsepower; a 4-speed transmission with hand-operated clutch; adjustable rear wheel tread; a belt pulley; power take-off; and an optional hydraulic lift mechanism that raised and lowered the tractor's implements.

The B Series underwent a number of modifications over 17 years that included: the introduction of larger 2-cylinder engines; a lengthened frame; a 6-speed transmission; electric starting and lighting; and remote hydraulics. The most striking change appeared in 1939, when a streamlined design was introduced. Fashioned by the noted industrial designer Henry Dreyfuss, the streamlined tractor is referred to as "styled", while the original design is referred to as "unstyled".

This volume in the Iconografix *Photo Archive Series* includes both unstyled and styled examples of the Model B Series tractors. This extraordinary collection illustrates the versatility of the Model B...the most popular tractor ever to bear the John Deere name.

The Unstyled Model B
1935 - 1938

JOHN DEERE MODEL B GENERAL PURPOSE TRACTOR
(Flywheel-Side View)

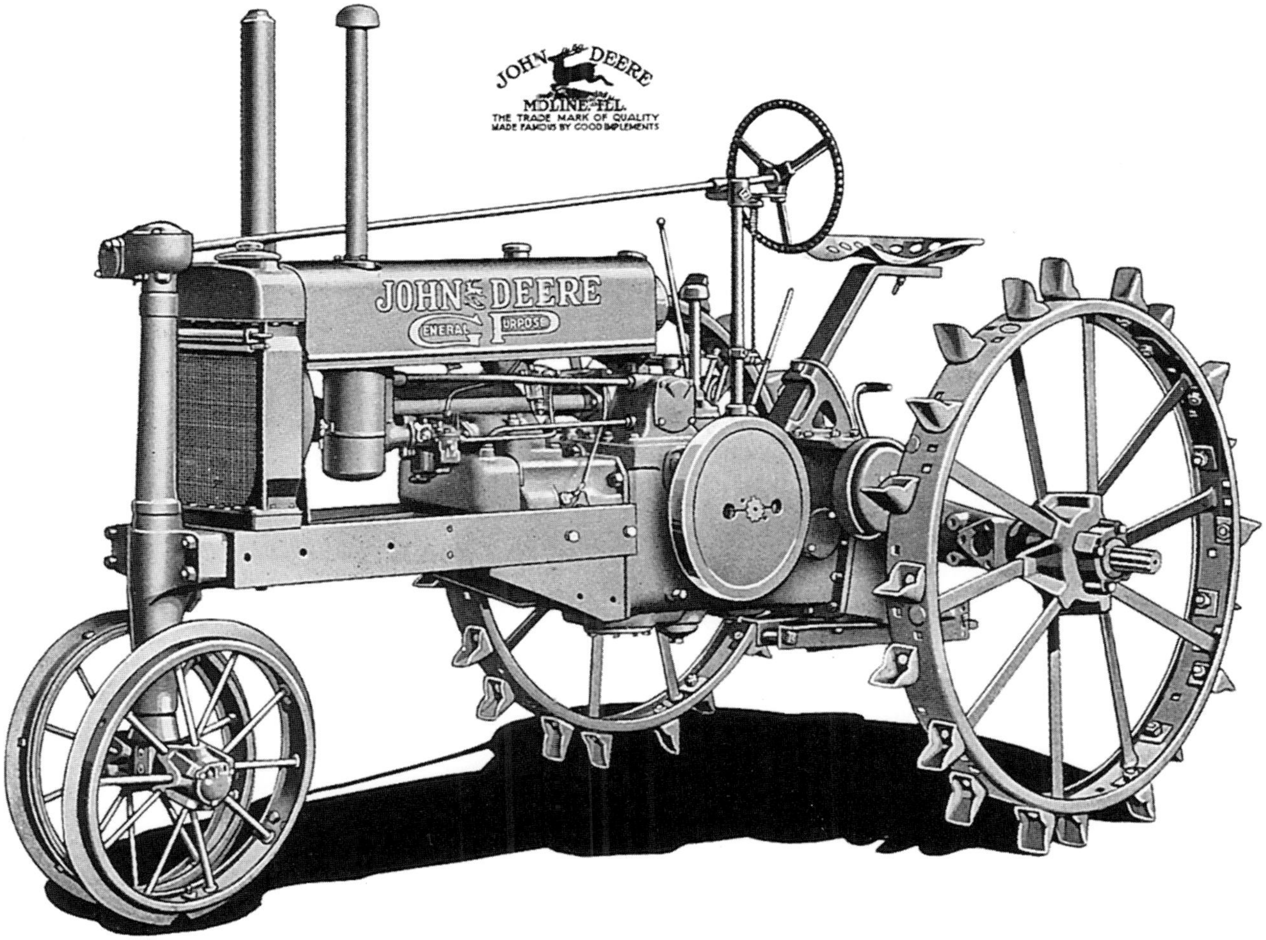

THESE MECHANICAL FEATURES TELL THE STORY OF JOHN DEERE SIMPLICITY, ECONOMY, DURABILITY

A parade of 45 new Model A and B tractors sold by Livingston County Tractor Sales. Pontiac, Illinois. February 1936.

View of flywheel side of an unstyled Model B on rubber tires.

The exposed radiator grille and steering pedestal characterized the unstyled Model B.

Rear view of unstyled Model B.

JOHN DEERE

John Deere-Lindeman crawler pulling 6-foot Killefer chisel plow. Conversion from Model BO orchard tractor by Lindeman Power and Equipment of Yakima, Washington. Photographed in Ashland, Oregon.

Model B with integral (mounted) 2-bottom plow. April 1935.

Hi-Crop version of Model BW, the Model BWH. Buena Park, California. August 1938.

Model B with B-100 series integral cultivator. Purdue University Soils and Crops Farm. Lafayette, Indiana. July 1935.

No. 10 corn picker and Model B. Flandreau, South Dakota. October 1936.

Model B with power mower. Geneseo, Illinois. June 1938.

Model B with 2-row integral cultivator. LaMoille, Illinois. June 1938.

Model F grain drill drawn by Model BO. August 1938.

Model AB15 potato hoe and Model B with hydraulic lift. August 1935.

Rake with experimental swing hitch mounted on Model B. Photographed outside factory. March 1938.

JOHN DEERE
Swing Hitch
3-7-1938

Model BW cultivating tomato plants near San Jose, California. July 1936.

Model B planting cotton with Model B37 bedder and planter. Alvarado, Texas. May 1937.

Model BW with B & B cultivator. Barker Miller Distributing Company. Phoenix, Arizona. June 1936.

Model BN with cultivator and vine rake. October 1937.

JOHN DEERE

Model B with Model 101 2-way plow. Lebanon, New York. May 1938.

Model BW and mounted cultivator. San Fernando Valley, California. August 1938.

Model B spraying potatoes. Troy, Ohio.

Model BW with bean sprayer. December 1937.

Model BN with cultivator. Phoenix, Arizona. November 1935.

Model B with 2-row mounted cultivator. Tapered hood allowed the operator an unobstructed view. LaMoille, Illinois. June 1938.

Model B and integral spring-tooth cultivator. 1937.

Model BN and Model B-614 lettuce cultivator. United Farms Co. Salinas, California. October 1936.

Model BO with disk harrow in orange grove. Buena Park, California. December 1935.

Model BO hauling apples. Berrien Springs, Michigan. October 1938.

Model B with drawn mower. June 1937.

No. 7 combine and Model B. July 1937.

Model B with Dain sweep rake and hay stacker.
Hammond, Wisconsin. August 1935.

Model B and power manure spreader.

MOLINE ILL. U.S.A.
POWER SPREADER

Model B and power manure spreader. Midvale Farm. May 1935.

JOHN DEERE
SPREADER

Portable grain elevator and Model B with dump wagon. Geneseo, Illinois.

Model B belt pulley powers No. 5 sheller. Jasper, Minnesota. January 1936.

JOHN
DEERE

Model B running irrigation pump. Grants Pass, Oregon. September 1937.

Model BI with 5-gang mower on Rock Island Arsenal golf course. July 1936.

Street flusher with Model BI. March 1936.

Model BI with Parsons Loadscoop.

JOHN DEERE

Model BI with Parsons Loadscoop. September 1937.

Integral 4-row cultivator with fertilizer attachment mounted on Model B. February 1938.

John Deere-Lindeman crawler shown with hydraulic tool bar and 6-foot dozer blade. Appears to be a page from a sales manual.

Row of 8 Model B tractors. August 1937.

Unstyled Model B with rubber tires and dog on seat. Norwalk, Ohio.

Unstyled Model B
Studio Artwork

Hi-Crop Model BNH.

Model BW.

Adjustable front and rear axles were features of the Model BW.

Orchard Model BO with citrus fenders.

Model BR with standard tread.

Model BN.

Model BO on rubber. September 1937.

Late Model BO with lights. November 1945.

Late Model BR. June 1945. (Editor's note: The BO and the BR remained in production until January 1947, but were never restyled.)

The Styled Model B
1939 - 1952

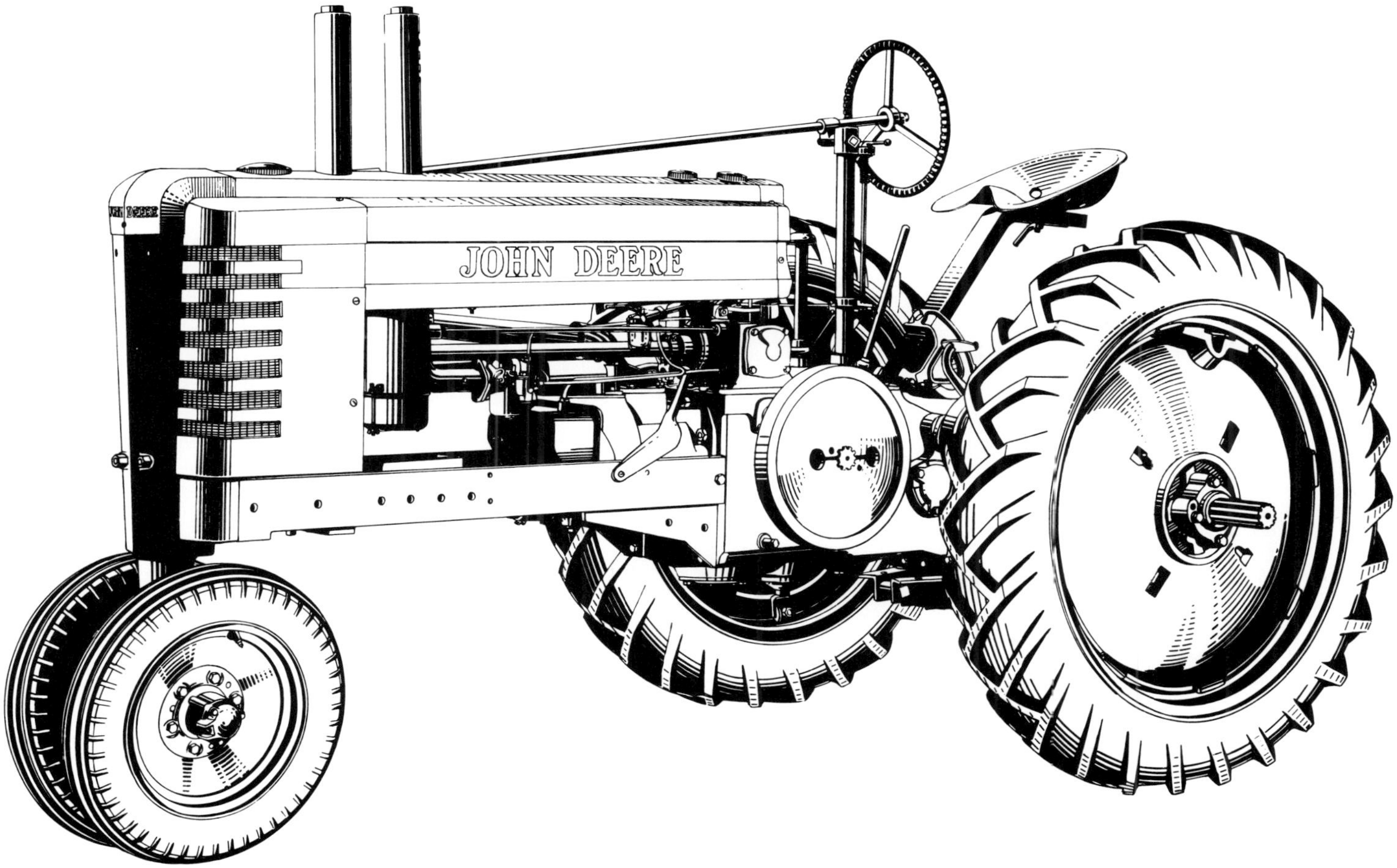
JOHN DEERE

Series of views showing styled Model B being loaded onto a trailer, pulled by Deere & Company truck, and being delivered and unloaded. Milan, Illinois. July 1939.

JOHN DEERE

DEERE & CO
43

JOHN

Model B restyled instrument panel. Styled tractor with electric starter and lights. September 1939.

View of electric starter mechanism. September 1939.

JOHN DEERE

Plowing at night by electric light. September 1939.

Model B with plow. Brandshaw, Maryland. May 1939.

Model B with Model A3 2-way plow. King's Farm. Moline, Illinois. August 1939.

Late Model B with Model JB double-action disk harrow. Experimental Farm, Coal Valley, Illinois. May 1948.

Model B with Ajax spike-tooth harrow. Deere Experimental Farm. May 1948.

Model B with 2-row planter. June 1939.

Model B with 2-row cultivator. Dubuque, Iowa. June 1951.

JOHN DEERE

Model 290 planter and Model B equipped with Powr-Trol remote hydraulics. June 1950.

Model B and plow. Cedar Rapids, Iowa. June 1938.

Model B with cultivator on Donalek Farm. Cedar Rapids, Iowa. June 1939.

No. B-253 cotton and corn planter and Model B. May 1939.

No. 15 cotton harvester mounted on Model B. Lubbock, Texas. December 1947.

Model B with cultivator in Texas cotton. July 1939.

Model B and beet lifter. Fort Collins, Colorado. November 1939.

Late styled Model B with side-delivery rake. Moline, Illinois. July 1951

Model B with Dain sweep rake. July 1939.

Model B with potato planter. Rockefeller Farm. Phelps, New York. June 1939.

Model B and 1-row, level-bed potato digger. Elbridge, New York. July 1940.

No. 12-A combine and Model B. August 1951.

No. 11 combine and Model B. May 1939.

JOHN DEERE

Model B with 16x18 pick-up hay press. The owner of this 200 acre farm in Morrow County, Ohio reported the work was completed with the aid of two 12-year old boys. August 1941.

Model B and Dain side-delivery rake. Clifton Spring, New York. August 1940.

Model B and side-delivery rake. Robinsville, Missouri. September 1938.

Model 116-W hay baler and late Model B. Muscatine, Iowa. June 1950.

Model B with No. 9 combine. Preebe Farm. Chamberlain, South Dakota. August 1939.

Model B with power mower and windrower in flax field. St. Ansgar, Iowa. August 1939.

Model B with power binder. Sibley, Iowa. July 1939.

No. 72 forage harvester and Model B. July 1951.

JOHN DEERE

JOHN DEERE
JOHN DEERE

Model B and Model 614 baler. August 1941.

No. 15 corn picker with Model B. November 1938.

Power corn binder with Model B. October 1938.

17917
FARM
TRAILER

Model B with Model 953 trailer. October 1946.

Model B and No. 11 combine. May 1939.

Model B with Model JBA disk harrow. Geneseo, Illinois. April 1951.

Model H spreader and Model B on Woods Farm. Coal Valley, Illinois. April 1948.

John Deere-Lindeman TZ-100 Plantrol transplanter operated by Model B. Independent spring pressure could be applied to the press wheels to obtain various degrees of packing around the plants. October 1947.

Styled Model B
Studio Artwork and Photography

No. 26 corn picker on Model B. December 1939.

Model BW. August 1938.

Model B with wheel removed to show detail of mounted B240 Lister.

Pulley-side view of Model B on steel wheels. May 1938.

Flywheel-side view of Model B on rubber tires. May 1938.

JOHN DEERE
JOHN DEERE

Newly styled Model BN. 1938.

Hi-Crop adjustable wide-front Model BWH. July 1939.

Hi-Crop narrow-front Model BNH. July 1939.

Model BN. July 1939.

Late BW with electric starting and lights, padded seat and pressed steel frame. October 1946.

Late BN. October 1946.

Late Model B. September 1946.

The Iconografix Photo Archive Series includes:

JOHN DEERE MODEL D Photo Archive	*ISBN 1-882256-00-X*	
JOHN DEERE MODEL B Photo Archive	*ISBN 1-882256-01-8*	
FARMALL F-SERIES Photo Archive	*ISBN 1-882256-02-6*	
FARMALL MODEL H Photo Archive	*ISBN 1-882256-03-4*	
CATERPILLAR THIRTY Photo Archive	*ISBN 1-882256-04-2*	
CATERPILLAR SIXTY Photo Archive	*ISBN 1-882256-05-0*	
TWIN CITY TRACTOR Photo Archive Volume I	*ISBN 1-882256-06-9*	*(Sept. 93)*
MINNEAPOLIS-MOLINE U SERIES Photo Archive	*ISBN 1-882256-07-7*	*(Sept. 93)*

The Iconografix Photo Archive Series is available from direct mail specialty book dealers and bookstores throughout the world, or can be ordered from the publisher.
For information write to:

Iconografix
P.O. Box 18433
Minneapolis, Minnesota 55418 USA